Corrosion Behavior of Austenitic Stainless Steels in CO2-Saturated Synthetic Oil Field Formation Water

Jorge Luiz Cardoso
Marcel Mandel
Lutz Krüger
Luís Flávio Gaspar Herculano
Pedro de Lima Neto
Marcelo José Gomes da Silva

ELIVA PRESS

ELIVA PRESS

Jorge Luiz Cardoso
Marcel Mandel
Lutz Krüger
Luís Flávio Gaspar Herculano
Pedro de Lima Neto
Marcelo José Gomes da Silva

The corrosion behavior of austenitic and superaustenitic stainless steels was investigated in carbon dioxide-saturated synthetic oil field formation water using cyclic polarization tests. In order to measure the effect of carbon dioxide pressure, the samples were also exposed in a pressurized medium containing carbon dioxide and synthetic air. For this purpose, tests were performed for long exposure time at 80 °C under 8 MPa of a mixture of carbon dioxide and synthetic air both acting together. The results indicated that the type of corrosion on the surfaces of the samples after all the tests was pitting corrosion. According to the results, the AL-6XNPLUS™ steel presented the best performance in all experiments. The results also indicated that the conventional austenitic stainless steels are not suitable for the use in carbon dioxide containing environment in aqueous medium.

Published by Eliva Press SRL
Address: MD-2060, bd.Cuza-Voda, 1/4, of. 21 Chişinău, Republica
Moldova
Email: info@elivapress.com
Website: www.elivapress.com

ISBN: 978-1-63648-061-9

Corrosion behavior of austenitic stainless steels in CO₂-saturated synthetic oil field formation water

Jorge L. Cardoso[a]*, Marcel Mandel[b], Lutz Krüger[c], Luís F. G. Herculano[d],
Pedro de Lima-Neto[e], Marcelo J. Gomes da Silva[f]

[a, d, f]Department of Metallurgical and Materials Engineering, Federal University of Ceará,Campus do Pici, bloco 729, Fortaleza 60440-900, Ceará, Brazil

[e]Department of Analytical Chemistry and Physical Chemistry, Federal University of Ceará

[b, c]Institute of Materials Engineering, Technische Universität Bergakademie Freiberg Gustav-Zeuner Str. 5, 09599, Freiberg, Germany

Previously published: CARDOSO, Jorge Luiz et al. Corrosion Behavior of Austenitic Stainless Steels in CO2-Saturated Synthetic Oil Field Formation Water. Materials Research, v. 22, n. 4, 2019.

Abstract

The corrosion behavior of austenitic and superaustenitic stainless steels was investigated in carbon dioxide-saturated synthetic oil field formation water using cyclic polarization tests. In order to measure the effect of carbon dioxide pressure, the samples were also exposed in a pressurized medium containing carbon dioxide and synthetic air. For this purpose, tests were performed for long exposure time at 80 °C under 8 MPa of a mixture of carbon dioxide and synthetic air both acting together. The results indicated that the type of corrosion on the surfaces of the samples after all the tests was pitting corrosion. According to the results, the AL-6XNPLUS™ steel presented the best performance in all experiments. The results also indicated that the conventional austenitic stainless steels are not suitable for the use in carbon dioxide containing environment in aqueous medium.

Keywords: stainless steel, cyclic polarization, pitting corrosion, carbon dioxide pressure

1

TABLE OF CONTENTS

1 INTRODUCTION

On the coast of Brazil, there is a salt layer that contains an oil of good quality located below it. This layer is called pre-salt (a geological formation on the continental shelves) and it is located under deep seawaters demanding good materials for extracting its oil[1]. The corrosion process in the pre-salt layer as well as in deep seawaters occurs under very specific conditions. The main characteristics under the pre-salt layer are high temperatures (between 80 °C and 150 °C), absence of oxygen, presence of gases such as carbon dioxide and hydrogen sulfide, high pressure, microorganisms and high content of dissolved salt[2]. Due to the high content of sodium chloride present in the salt layer, localized corrosion in this environment is very common and it is the most difficult process to control. The oil and gas industry are concerned about the impact of oil accidents on marine ecosystems. These accidents can be avoided by using materials more and more corrosion resistant in such conditions. Since 1970, corrosion resistant alloys have been used in oil extraction from very deep marine oil fields instead of the cheap carbon steel pipelines due to the difficulty attending the addition of corrosion inhibitors in carbon steel[3]. Currently, there is a great interest, here in Brazil, to study the corrosion resistance of stainless steels with higher chromium content in their composition, such as the super austenitic stainless steels, because it is expected that such stainless steels are more corrosion resistant than the conventional stainless steels with 12 wt.% Cr.

Some studies about the corrosion resistance of austenitic stainless steels in CO_2 containing environments were already published[4-8]. Russick et al[4] studied the corrosion mechanism for several materials and, among them, 316 steel and C1018 carbon steel in supercritical carbon dioxide containing environment. The authors showed that significant corrosion occurs only for the C1018 carbon steel. Furukawa et al[5] investigated the corrosion resistance in supercritical carbon dioxide pressurized at 20 MPa between 400 °C and 600 °C. The authors stated that no effect of carbon dioxide pressure on corrosion behavior was observed for both alloys. In another research involving the carbon dioxide gas[6], some results indicated that the protective layers on the tested steels, among them AL-6XN steel, were the protective Cr-rich oxide phases such as Cr_2O_3 and $Cr_{1.4}Fe_{0.7}O_3$. The formation of oxide layers was identified as the primary corrosion mechanism on the AL-6XN under high pressure of carbon dioxide of 20.7 MPa at 650 °C and the effect of alloying elements such as aluminum, molybdenum, chromium and nickel was also suggested[7]. G. Cao et al[8] studied the corrosion behavior of some stainless steels (310, 316 and 800H steels) under high pressure of carbon dioxide of 20 MPa at 650 °C and they identified three major peaks for the surface X-ray diffraction: (i) the austenitic peak that represents the base alloy, (ii) magnetite (Fe_3O_4) and spinel phase peaks corresponding to $(Cr, Mn, Fe)_3O_4$ oxides, and (iii) chromium rich oxides of $Cr_{1.3}Fe_{0.7}O_3$ or Cr_2O_3. For 316 steel, the peaks found were identified as Fe_3O_4, $FeCr_2O_4$ and peaks of austenite. They also concluded that for 316 steel, the spinel oxide ($FeCr_2O_4$) layer is less protective than the chromium-rich ones ($Cr_{1.3}Fe_{0.7}O_3$ or Cr_2O_3) for 310 steel.

From the viewpoint of corrosion resistance, it is expected that super austenitic stainless steels have better performance in aggressive environments when compared with the 300 series of austenitic stainless steels. The corrosion behavior of austenitic stainless steels (304L, 316L and 317L) and super austenitic stainless steel (AL-6XNPLUS™) was recently studied by the authors of this work[9].

Thus, in this paper, the corrosion behavior of a super austenitic stainless steel (AL-6XNPLUS™) in CO_2-containing environment was studied and compared with the corrosion behavior of two conventional austenitic stainless steels (316L and 317L) for the same conditions using electrochemical techniques and pressurized experiment with CO_2 and synthetic air.

2 MATERIALS AND METHODS

2.1 Material

The materials used in this study were the AL-6XNPLUS™ super austenitic stainless steel and two conventional austenitic stainless steels (316L and 317L). The materials were received in plates containing an average thickness of 0.30 cm. Table 1 presents the chemical composition of the steels measured in an Optical Emission Spectrometer (PDA-7000 SHIMADZU) and their Pitting Resistance Equivalent Number (PRE_N) calculated according to equation 1[9].

$$(PRE_N) = \%Cr + 3.3 \,\%Mo + 30 \,\%N \tag{1}$$

Table 1. Chemical composition and pitting resistance equivalent number (PRE_N) of the studied alloys (wt%).

Alloys	C	N	Mn	Si	Cr	Ni	Mo	PRE_N
316L	0.030	0.05	1.65	0.41	17.2	10.7	2.2	26
317L	0.024	0.06	1.49	0.40	17.8	12.3	3.5	31
AL-6XN PLUS™	0.021	0.24	0.35	0.32	21.8	25.8	7.6	54

2.2 Experimental methods and equipments

For the metallurgical characterization of the samples in the as received condition, X-ray diffraction (XRD) using Synchrotron Light (energy 12 keV) was used to detect the phases presented in these steels. The samples were manufactured according to TMEC Project – Gleeble. The shape and dimensions of the samples (in mm) used for this characterization are shown in Figure 1. The samples were fixed inside the Gleeble as shown in Figure 2. No stress was

5

applied on the samples. A database called Joint Committee for Powder Diffraction Data (JCPDS) belonging to ICDD database (International Centre for Diffraction Data) was used to identify the peaks. These measurements were carried out at the Brazilian Synchrotron Light Laboratory in the city of Campinas-SP in Brazil.

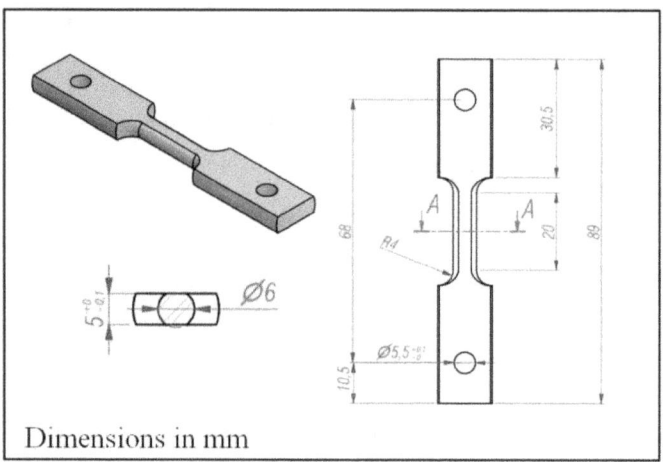

Fig. 1. Shape and dimension (in mm) of the samples used for the XRD measurements.

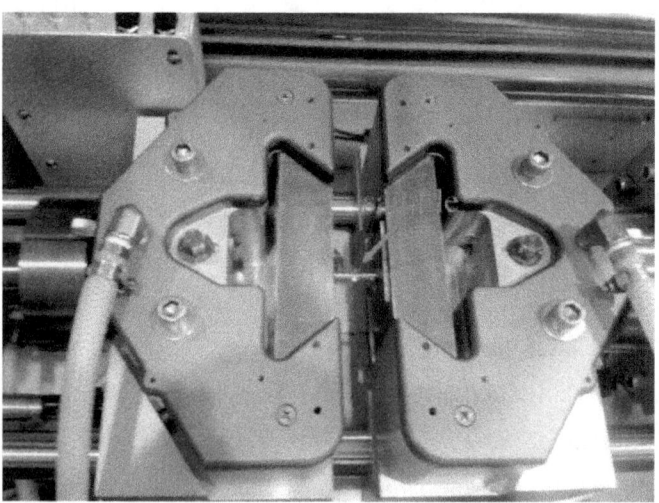

Fig. 2. Photograph of the sample fixed inside the Gleeble for the XRD measurements.

As sigma phase (σ) is one of the most harmful phases for stainless steels, a simulation of the sigma peaks for austenitic stainless steels using the PowderCell software was carried out. The result is shown in Figure 3. From the simulation, the sigma phase peaks appeared between 25° and 35° (2 theta). The XRD simulation was carried out for a wavelength of 0.10332 nm.

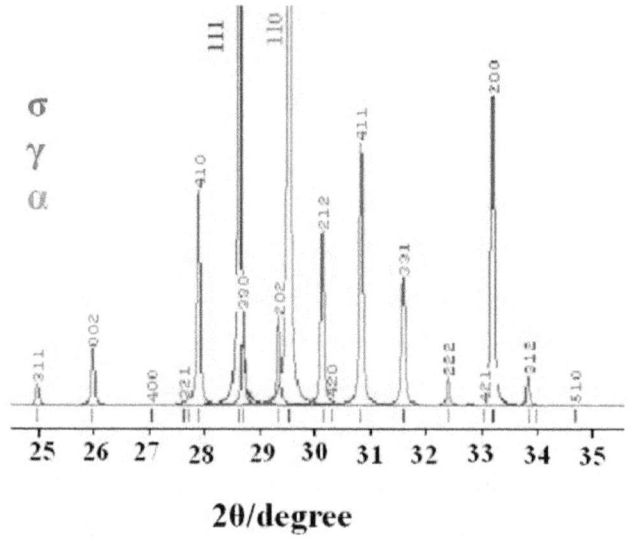

Fig. 3. Sigma peaks simulation for austenitic stainless steel.

The electrochemical measurements were carried out at room temperature using the cyclic polarization technique. In the preparation for the electrochemical tests, the samples were mounted in cold curing epoxy resin, ground with 600 grade sandpaper, rinsed with ethanol and dried before each measurement. The dimensions of the samples were on average 8.0 mm x 8.0 mm x 3.5 mm. The samples were coated with a lacquer to reduce crevice corrosion on the epoxy/steel edges leaving an average exposed area of 39 mm². All the samples were investigated in the as received condition. The electrochemical experiments

7

were performed in a single-compartment Pyrex® glass cell with a Teflon cover containing holes to fix the electrodes. A saturated Ag/AgCl and a platinum foil of 2 cm² were used as the reference electrode and the counter electrode, respectively. A CO_2-saturated synthetic oil field formation water was used as electrolyte with the chemical composition shown in Table 2.

Table 2. Chemical composition of the used electrolyte calculated for 1 L of distilled water.

Reagents	$CaSO_4$	$MgCl_2$	$NaHCO_3$	NaCl
C (g/L)	0.516	4.566	0.425	29

A potentiostat (AUTOLAB PGSTAT302N) controlled by a computer using the software Nova 11.1, which allowed the acquisition and analysis of the electrochemical data was used. Before the electrochemical tests, the electrolyte was deaerated with nitrogen to simulate pure oxygen-free environment below the pre-salt layer. Upon reaching a pH of approximately 8.2 ± 0.1, the solution was deaerated. Afterwards, the nitrogen flow was decreased and the solution was flushed with carbon dioxide until the pH was stabilized at 5.1 ± 0.1, indicating the saturation with carbon dioxide. The final pH of the solution was acid. After saturation, the samples were immersed into the solution and the system was maintained in open circuit potential (E_{oc}) for 30 minutes. In the next step, the cyclic polarization tests were started and the potential was swept with a scan rate of 0.33 mV s^{-1} from -0.5 V up to 1.2 V in relation to the E_{oc}. Electrochemical tests were also carried out without flushing CO_2 in order to compare the results. After the cyclic polarization tests, the samples were cleaned with water, rinsed with ethanol and dried. Micrographs of the surface of the samples after electrochemical tests were obtained using a scanning electron microscope (SEM Philips XL-30). The electrochemical tests were repeated three times to ensure reproducibility.

8

For the pressurized tests, the samples were cut in sheets with the following dimensions: 3.2 cm x 6.6 cm x 0.19 cm (316L), 3.0 cm x 8.2 cm x 0.30 cm (317L), and 3.2 cm x 8.2 cm x 0.59 cm (AL-6XN PLUS™). The pressurized corrosion tests were carried out in a system comprised of the following components: gas supply system, a thermostat (BTC-3000) and a high-pressure laboratory reactor (BR-300, 1.4571). Two gases, synthetic air (80% Vol.N_2; 20% Vol.O_2) and carbon dioxide gas (99.995% of purity) were used. In order to evaluate the effect of the gases on the corrosion of the samples, the gases were mixed (62.5% CO_2 and 37.5% synthetic air). The total pressure used was 8 MPa (5 MPa of CO_2 and 3 MPa of synthetic air) at 80 °C. This was the maximum pressure that could be applied in the autoclave with safety. Figure 4 depicts the samples on the specimen holder and inside the autoclave.

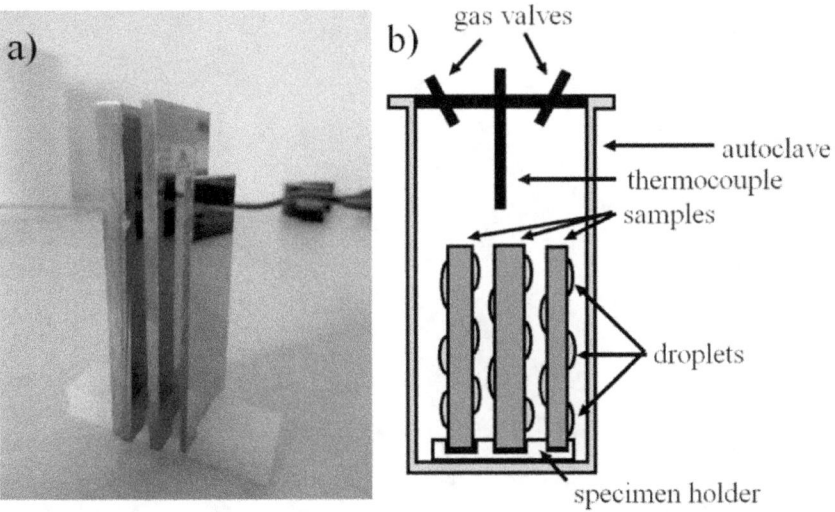

Fig. 4. Samples fixed on the specimen holder (a) and a schematic illustration of the samples positioned inside the autoclave (b).

Before the tests, the samples were cleaned with distilled water, rinsed with ethanol and dried. Afterwards, the surfaces of the samples were sprayed with

9

the synthetic oil field formation water solution (see Table 2). The experiments were performed for 168 h. After the exposure tests, the samples were analyzed using an optical microscopy for identification of degradation and corrosion product formation. Afterwards, the samples were cleaned with HCl 20% to remove the corrosion products and their surfaces analyzed by scanning electron microscopy (TESCAN, MIRA3 XMU9). The corrosion products in the shape of powder were also analyzed by XRD ($Cu_{K\alpha}$, λ = 0.15406 nm). The FIZ/NIST Inorganic Crystal Structure software, version 1.9.5, was used to identify the peaks. Finally, the surface topography was investigated by a confocal white light interferometric surface measuring system (SMS *Oberflächen-Messsystem, Breitmeier Messtechnick*). The corroded surfaces of the samples were used to evaluate the depth of the localized corrosion attack. All experiments were carried out in duplicate. The pressurized tests were carried out at Freiberg University of Mining and Technology in Germany.

3 RESULTS AND DISCUSSION

3.1 Characterization of the materials in the as received condition

Figure 5 shows the X-ray diffractogram pattern for the 316L steel. The angle 2θ was measured between 25° and 79° to detect the main phases. A synchrotron light radiation source ($\lambda = 0.10332$ nm) was used. Austenite peaks (FCC) and some ferrite peaks (BCC) on the diffractogram pattern for the 316L steel were observed as seen in Figure 5. The same result was found for the AL-6XNPLUS™ steel as can be seen in Figure 6. The results showed that the two materials are not solution annealed because of the presence of ferrite phase. No sigma phase peaks were detected between 25° and 35° as predicted by the simulation in Figure 3. This means that the results involving reduction of corrosion resistance cannot be attributed to this phase. These measurements were not possible for the samples of the 317L steel. Due to a manufacturing problem, the samples of this steel did not fit in the Gleeble.

Figure 7 shows the microstructure of the AL-6XN PLUS™ super austenitic stainless steel (the main steel of this research) in the as received condition after an electrolytic etching with oxalic acid 10 %. It is possible to see the grain boundaries and the twin boundaries. This is a characteristic of the austenitic phase.

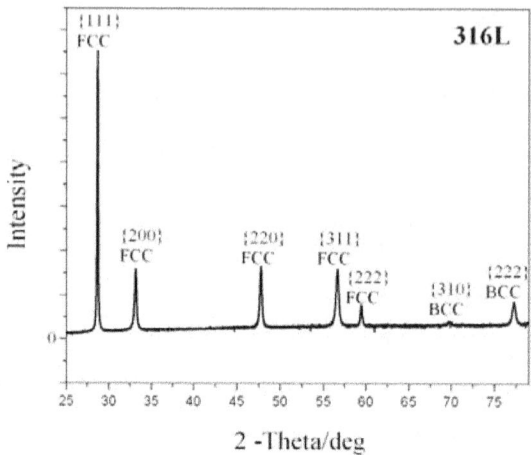

Fig. 5. XRD pattern for the 316L steel in the as received condition (synchrotron light radiation source, $\lambda = 0.10332$ nm).

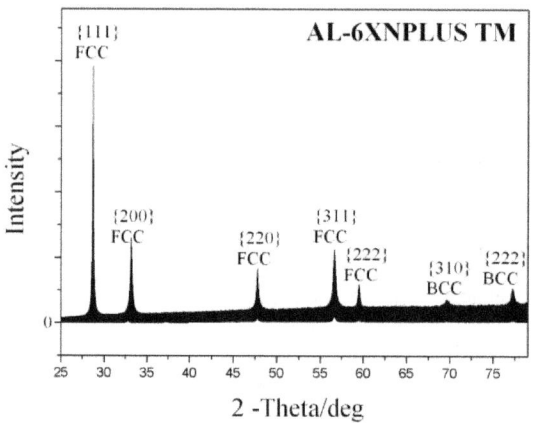

Fig. 6. XRD pattern for the AL-6XNPLUS™ steel in the as received condition (synchrotron light radiation source, $\lambda = 0.10332$ nm).

Fig. 7. SEM image of the microstructure of the AL-6XN PLUS™ super austenitic stainless steel.

3.2 Cyclic Polarization tests

The potentiodynamic cyclic polaritation curves (in their linear and semilogarithmic form) for the measurements in CO_2-saturated synthetic oil field formation water are given in Figure 8.

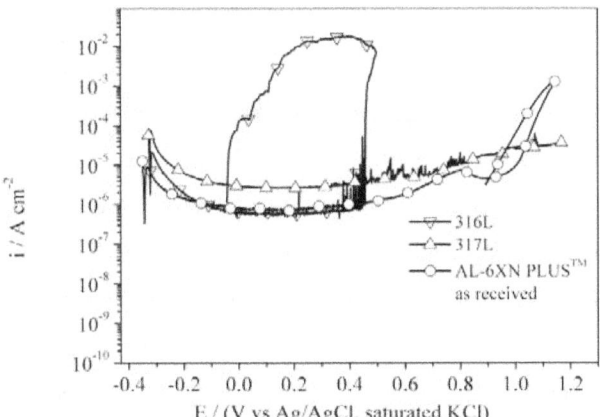

Fig. 8. Cyclic polarization curves for the samples in CO2-saturated synthetic oil field formation water.

It can be noticed the formation of passive layers in the direct scan of all investigated samples. For the super austenitic steel (AL-6XNPLUS™), the direct scan showed a small peak about + 0.80 V, which was attributed to the breakdown of the passive film, followed by repassivation of the steel surface, and for the applied potentials higher than +0.96 V there was an increase in current density with the applied potential, which indicated that the transpassive region was reached. The passive region is quite stable. The reverse scan showed a small loop indicating no localized corrosion. The increase in current density after the potential of +1.01 V on the cyclic polarization curves for the super austenitic steels is associated with water dissociation (oxygen evolution) according to equation 2[10]. With the release of the oxygen gas from the water molecule, there is the continuation of the oxidation process on the surface of the sample. According to Bandy & Cahoon[10], with this type of reaction occurring, it is impossible to distinguish the current due to the metal corrosion from the current of the water dissociation leaving the electrochemical tests limited for very high potentials (above +1.0 V).

$$2H_2O \rightarrow O_2 + 4H^+ + 4e^- \tag{2}$$

For the 317L steel, no hysteresis was observed on its cyclic polarization curve. The reverse current returns over itself suggesting that there was no localized corrosion. The cyclic polarization curve for the 316L steel also presented the same passivation behavior in the beginning as the passivation of the super austenitic steels (order of magnitude of 10^{-7} A / cm²), however, between the potentials +0.36 V and +0.45 V, the curve had a noise indicating fragility of the passive film. In high chloride concentration solutions, the pit is characterized by a minimum potential, called pitting potential. Below this potential, the metal remains passive and, above it, pits are formed, which is a criterion used for their

14

detection, although a detailed examination of the passive region shows that the passivation current is noisier in chloride solutions than in solutions in which this ion is absent. This effect can be seen in Figure 8 for the alloy 316L. After reaching the potential of +0.45 V (pitting potential), the passive film of the alloy 316L was broken and there was a sudden increase in current density with high values (order of magnitude of 10^{-3} A / cm^2). The breakdown of the passive film for austenitic stainless steels occurs in the presence of chloride ions which subsequently results in the initiation of pits by an autocatalytic process in which there is a local increase in chloride and acid concentrations due to corrosion product hydrolysis in cavities[11]. The reverse curve showed hysteresis indicating pitting formation on the surface of the 316L steel. The hysteresis curve closed at the potential of -0.039 V. Therefore, the formation of a positive hysteresis showed that the 316L steel did not obtain a good CO_2 corrosion resistance in the electrolyte used.

Malik et al[12] studied the relationship between pitting potential and Pitting Resistance Equivalent Number (PRE_N) of some stainless steels (austenitic, ferritic and duplex) at 50 °C in Gulf seawater under salt spray conditions and corrosion rates were determined by applying the electrochemical polarization resistance technique. Their results indicated that the presence of alloying elements such as chromium, molybdenum and nickel have a significant and beneficial influence on the pitting and crevice corrosion resistance of stainless steels. For stainless steels, this relationship is given by the empirical equation 1 shown before[9,13]. The higher the PRE_N, the better the pitting corrosion resistance is.

Although not presented in equation 1, nickel has also an important role in improving the corrosion resistance of stainless steels[14]. The super austenitic stainless steels contain in their composition chromium, molybdenum, nickel and nitrogen enough to guarantee a good performance of the passive film. This effect can be seen by the decrease of the anodic current with time on their polarization curves (passive region).

The difference between these materials (austenitic and superaustenitic) is in the composition. The more contents of alloying elements a steel possesses, more resistance against several forms of corrosion it is. An example are the elements chromium and molybdenum. These elements can adhere on the passive film to inhibit localized corrosion. An oxide layer of chromium and molybdenum can form on the surface of these steels and this layer can block the action of chloride ions by inhibiting the formation or growth of the pits. The element molybdenum on the passive layer can also change the electronic properties reversing the ion selectivity in the film structure hindering the migration of chloride ions through the film[15]. Molybdenum gives greater resistance to localized corrosion by forming molybidates incorporating on the passive film to improve its structure and also reinforces the passive film by increasing its thickness[16].

According to Sedriks[17], on a polarization curve, the greater the difference between the breakdown potential and the corrosion potential ($\Delta E = E_b - E_{corr}$), more resistant to corrosion the material is. Table 3 shows the values for the corrosion potential (E_{corr}), breakdown potential (E_b) and the difference between them (ΔE) calculated from the polarization curves of the studied alloys. The ΔE range is higher for the super austenitic stainless steel which confirms its high performance in relation to carbon dioxide corrosion. The 317L steel had a similar behavior like the super austenitic ones. The 316L steel showed the lowest value for ΔE indicating not being a suitable material for applications requiring good carbon dioxide resistance.

Table 3. Table with potential E_{corr}, E_b and ΔE in V (Ag/AgCl).

Alloys	E(corr)	E(b)	ΔE
316L	-0.32	+0.45	0.77
317L	-0.38	+0.99	1.37
AL-6XN PLUS™	-0.34	+0.98	1.32

The micrographs of the samples obtained after the end of the cyclic polarization measurements in CO_2-saturated synthetic oil field formation water are shown in Figure 9. One can see clearly the pits formed on the surface of the 316L steel. This explains the appearance of hysteresis on its polarization curve. The pits on stainless steels are generally spaced apart and most of the surface is passive. However, the pitting propagation speed is very fast[18, 19].

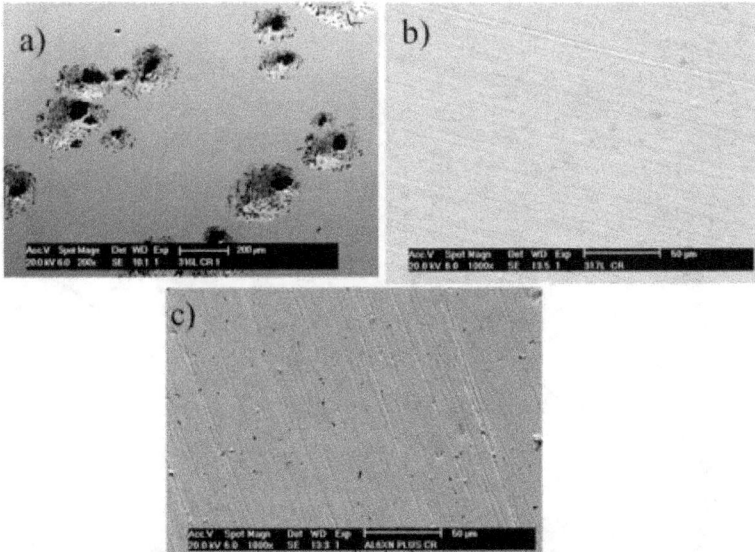

Fig. 9. SEM image of the alloys surface in the as-received condition after the cyclic polarization tests in CO_2-saturated aqueous medium. a) 316L, b) 317L, and c) AL-6XN PLUS™.

The super austenitic steel (AL-6XN PLUS™) showed no pits on its surface. This result is in agreement with its cyclic polarization curve with the absence of hysteresis. The pits formed on the surface of the 316L steel are not uniform and

17

their tendency is to grow even more over time. Figure 10 shows a specific pit on the surface of the 316L steel. One can see the total destruction of the material in the center of the pit and around the center, other micro-pits in growth state can be seen. The direction of pitting growth is from the center to the edges. The pits formed on the surface of the 316L steel sustain by themselves perforating the material. With the breakdown of the passive layer, an electrolytic cell is formed. The cathode region is the passive layer while the anode is the exposed metal, more precisely, the center of the pit. The flow of electrons between the anode and cathode is due to a large potential difference between these two regions. The corrosion process in this case is accelerated into the pit[17].

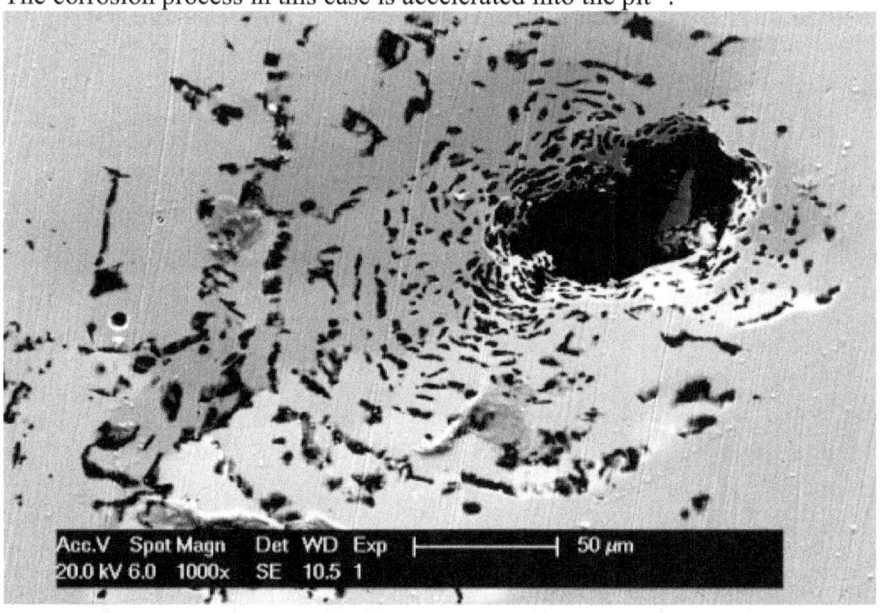

Fig. 10. SEM image of a specific pit on the surface of the 316L steel after the cyclic polarization tests in CO_2-saturated aqueous medium.

Figure 11 shows the cyclic polarization curves for the steels in the as-received condition in an aqueous medium of synthetic oil field formation water. The solution was deaerated again with nitrogen, but not saturated with CO_2. The solution was basic (pH = 8.1). All the samples suffered passivation with low current densities. The polarization curve for the sample of the 316L steel showed again hysteresis indicating localized corrosion. The absence of CO_2 in the solution caused the displacement of the pitting potential to more noble direction (more positive) leaving the alloy more resistant to localized corrosion. However, at the potential of +0.73 V there was the breakdown of the passive film followed by a high increase of the current density and subsequent formation of pits on the surface of the sample. The other steels (AL-6XN PLUS™, and 317L) showed a good corrosion resistance in the aqueous medium of synthetic oil field formation water. There was no hysteresis formation on their cyclic polarization curves, therefore, no pitting formation.

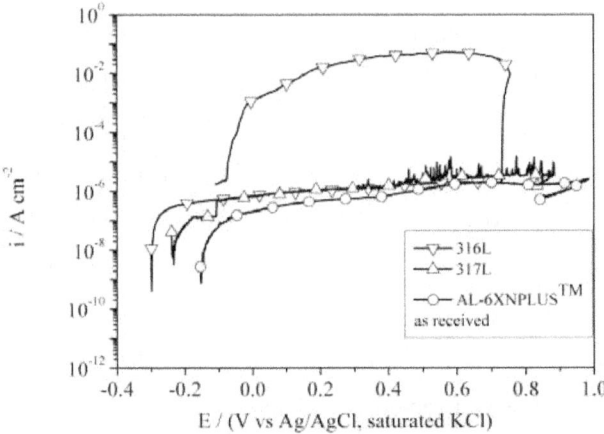

Fig. 11. Cyclic polarization curves for the alloys in the as-received condition in aerated synthetic oil field formation water without bubbling CO_2.

The pH of the solution is influenced by carbon dioxide gas. The corrosion rate tends to be lower when the solution is basic. When bubbling the solution with carbon dioxide gas, the following reactions occur:

$$CO_{2(g)} + H_2O_{(aq)} \leftrightarrow H_2CO_{3(aq)} \tag{3}$$

$$H_2CO_3 \text{ (aq)} \leftrightarrow 2H^+ \text{ (aq)} + CO_3^{2-} \text{ (aq)} \tag{4}$$

$$Fe(s) \leftrightarrow Fe^{2+} \text{ (aq)} + 2e^- \tag{5}$$

$$CO_3^{2-} \text{ (aq)} + Fe^{2+} \text{ (aq)} \leftrightarrow FeCO_3 \text{ (s)} \tag{6}$$

First, the carbon dioxide (CO_2) gas in the solution reacts with water forming carbonic acid (H_2CO_3), according to equation 3[20]. The pH of the solution changes from basic to acid values due to the dissociation of carbonic acid presented in the solution releasing H^+ ions, which leaves the solution more corrosion aggressive (equation 4). In addition, the carbonate ion reacts with Fe^{2+} to form iron carbonate ($FeCO_3$), as shown in equation 5 and 6. For stainless steels, this reaction can also occur with other elements such as chromium and molybdenum forming $CrCO_3$ or $Mo(CO_3)_2$ [21]. Figure 12 shows a comparison of pit density on the surface of the 316L steel in the as-received condition in CO_2-saturated aqueous solution (Figure 12a) and without CO_2 (Figure 12b). In the first case, the pit density on the surface of the 316L steel is greater than the pit density for the second case. This result shows the influence of the pH of the solution. Chloride is more aggressive in acid environment.

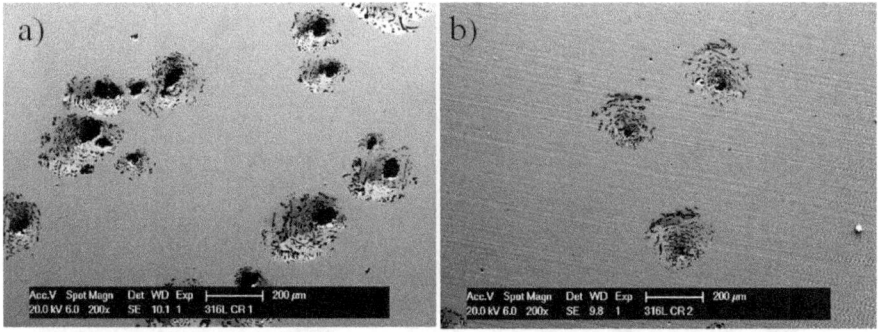

Fig. 12. SEM image of the pit density for the alloy 316L in the as-received condition in an aqueous medium (oil field formation water) a) with CO_2 and b) without CO_2.

The pitting potential is a function of the medium composition, concentration of aggressive ions, temperature, alloy composition and the surface treatment. As seen in Table 4, the pitting potential (E_p) is greater in basic solutions than in acid ones as expected. It seems that the pH of the solution had also influence on the size and density of the pits as shown in Figure 12.

Table. 4. change of the pitting potential and the corrosion potential measured in V vs Ag/AgCl sat KCl of the alloy 316L.

Alloy 316L			
pH	condition	Ep	Ecorr
5.2	as-received	+0.45	-0.32
8.1	as-received	+0.72	-0.30

3.3 High pressure tests

After the exposure tests under 5 MPa of CO_2 and 3 MPa of synthetic air at 80 °C for 10080 min, the samples were examined by optical microscope. Corrosion product formation (red rust) was observed on the surface of the 316L and 317L steels (black arrows) as shown in Figure 13.

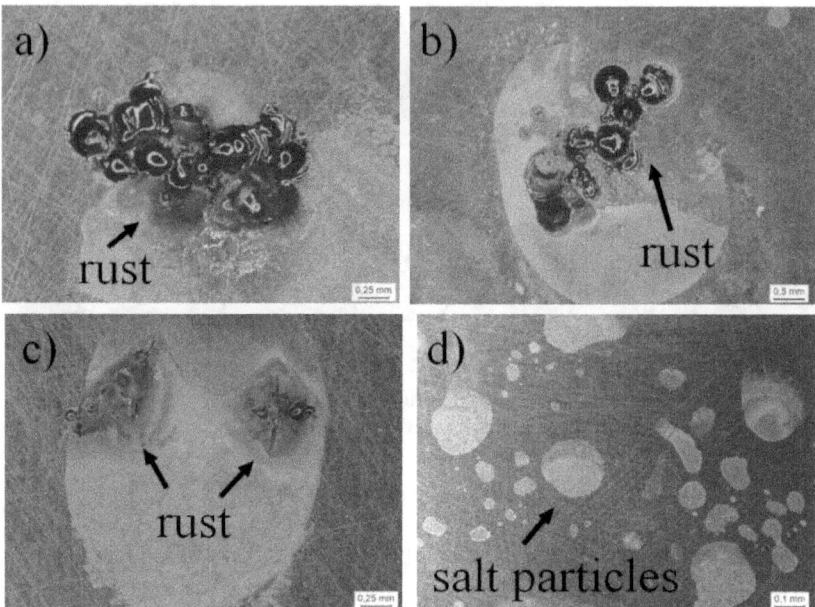

Fig. 13. Optical images of rust on the surfaces of the samples of the 316L steel (a, b), 317L steel (c) and salt particles on the surface of the AL-6XN PLUS™ steel (d) after the exposure test under synthetic air pressure of 8 MPa at 80 °C for 168 h sprayed with TQ3219 solution.

The rust can be considered as the final process of the corrosion process and it is located inside the region where there were droplets on the surface of the samples. Corrosion products are primarily composed of $Cr_{1.3}Fe_{0.7}O_3$ or Cr_2O_3 and $FeCr_2O_4$ that was identified by XRD ($Cu_{K\alpha}$, $\lambda = 0.1540$ nm) as shown in Figure 14. Rust on the surface of the AL-6XNPLUS™ steel was not found but some particles of salt were detected. It can be seen that the droplets of the solution act as anodic region and the sites around the droplets act as cathodic region.

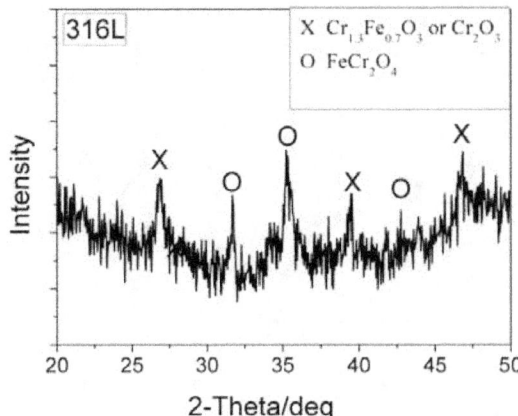

Fig. 14. XRD patterns of the corrosion product of the 316Lsteel after exposure tests (Cu$_{K\alpha}$, λ = 0.1540 nm).

Figure 15 shows the SEM of the surface of the 316L and 317L steels after the exposure test under CO_2 and synthetic air pressure (8 MPa at 80 °C for 10080 min) and after removing of the corrosion products. The kind of corrosion on the surface of the 316L and 317L steels was identified as pitting corrosion. The 316L steel was the most damaged steel comparing to the super austenitic one. Here there is a combination of factors that resulted on pitting corrosion: presence of chloride, presence of oxygen, CO_2 and synthetic air pressure, temperature and exposure time. The pressure acting on the surface of the samples compresses the solution against their surfaces enabling more effective action of chloride ions. As the samples were sprayed with the solution, there were sites on the surface with more solution (droplets) than others. The presence of pits was detected in sites of the surface where there were droplets of the solution. The chemical reactions that favored the pits formation occurred within the droplets (the combination of the two gases and the solution).

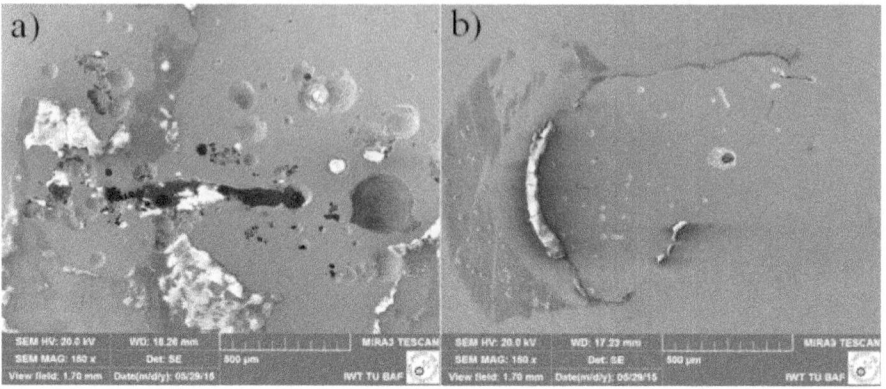

Fig. 15. Pits and salt particles on the surface of the sample of the 316L (a) and 317L (b) steels after removing the corrosion products of the pressurized test.

The topographies of the 316L and 317L steels are shown in Figure 16 and Figure 17. These results are in accordance with the SEM images and the cyclic polarization curves. The deepest pit found on the surface of the 316L surface possesses 0.20 mm and for the 317L steel, 0.026 mm.

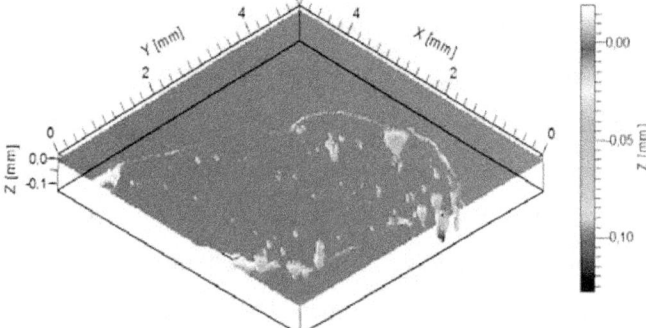

Fig. 16. Topography of the 316L steel showing the depth and the distribution of the pits after pressurized tests.

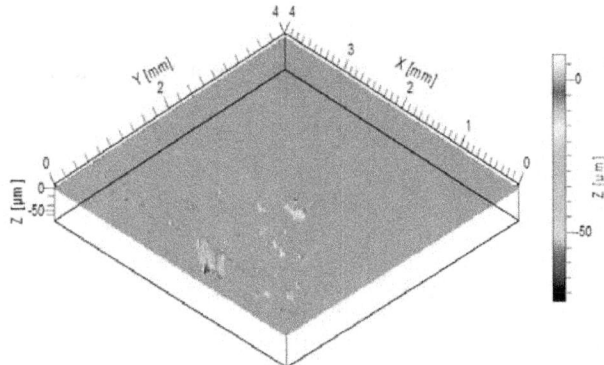

Fig. 17. Topography of the 317L steel showing the depth and the distribution of the pits after pressurized tests.

Regarding corrosion in aqueous medium, water plays a role as a solvent, dissolving gases providing several of the cathodic reactions for corrosion to occur. All the electrochemical reactions occurred inside the droplets. The chloride ions of the synthetic oil field formation water presented within the droplets reacted with the surface of the metal breaking the passive layer causing pits. Austenitic steels are iron based alloys and equation 7 presents a possible anodic reaction inside the pit after the breakdown of the passive layer for iron based alloys[22].

Inside the pit occurs the following anodic reaction (dissolution of iron)

$$Fe(s) \leftrightarrow Fe^{2+}(aq) + 2e^- \tag{7}$$

In the cathodic reaction, electrons flow to the cathode to be discharged. This occurs on the passive layer according to equation 8[23].

$$\frac{1}{2} O_2 + H_2O + 2e^- \leftrightarrow 2(OH^-) \tag{8}$$

As a result of these reactions, the charge inside the pit is positive and the charge surrounding the pit is negative. The positive charge into the pit (Fe^{2+}) attracts

25

the negative ions of chloride (Cl⁻) and this increases the activity into the pit according to equation 9[22].

$$FeCl_2 + 2H_2O \leftrightarrow Fe(OH)_2 + 2HCl \tag{9}$$

Due to formation of HCl, the pH inside the pit decreases which causes further acceleration of pitting corrosion.

Schematic drawings were made to depict the mechanism of the pitting corrosion for the exposure tests (Figure 18). Figure 18a depicts the pit initiation inside the droplet. The negative ion (Cl⁻) breaks the passive film. This penetration mechanism involves the migration of aggressive Cl⁻ ions from the solution through the passive layer under the influence of pressure and temperature. The breakdown of the passive film starts when cracks appear in the passive film under induced corrosion activity. This is enough to expose small areas on the surface of the metal to the solution. The cations from the metal are transferred from the passive film to the solution. This leads to the dissolution of the metal causing the thinning and final removal of the passive layer. The pits initially grow in the metastable condition[23]. Figure 18b depicts the next stage of pit growth on the surface of the metal. The pit becomes deeper with time while the droplet dries. A corrosion product (rust) forms on the surface of the metal and becomes thicker with time. The rust on the surface of the samples was identified as a chromium rich oxide (Figure 14). According to Rothman et al[23], this is typical for Fe-Cr-Ni stainless steels due to the greater diffusion coefficients of chromium and iron. According to equation 9, inside the pit there is a formation of HCl leaving the pH acidic within the pit. This accelerates the corrosion process in the bottom of the pit. Figure 18c depicts the last stage of pit growth on the surface of the metal. The Cr-oxide layer covers all the pit blocking the diffusion of Cl⁻ into the pit. The pit is stabilized. The Cr-oxide layer may be protective or not. The droplet is nearly dried reducing the moisture and the possibility for new pits to grow. Only the pressure and temperature would not be enough to cause this kind of corrosion on the surface of the samples. An aqueous medium is necessary and it seems to be the driving force for corrosion to occur.

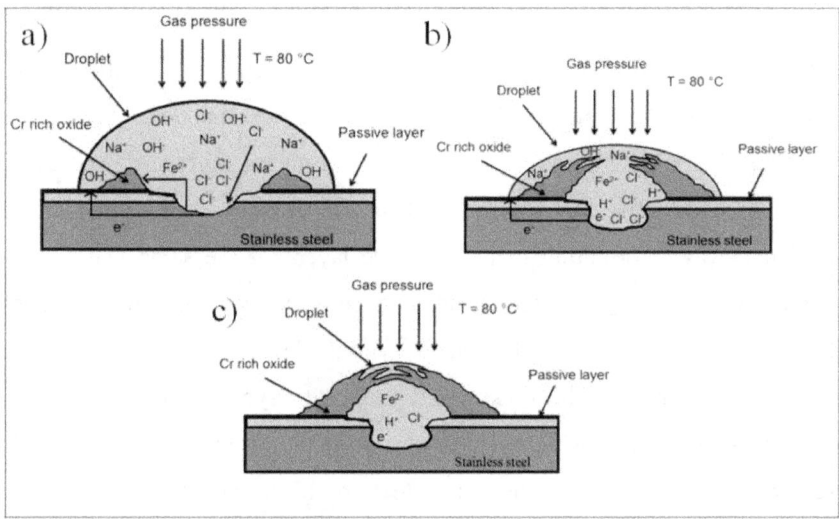

Fig. 18. Schematic illustration of the mechanism of pitting initiation (a), growth (b) and final stage (c) on the surface of the 316L steel.

Austenitic stainless steels are really good materials regarding to corrosion resistance. The results showed that there is no corrosion out of the droplets meaning that moisture has an important role on the corrosion process. Only the gases, no matter the combination between them, without an aqueous medium would not be enough to cause damage on the surface of the samples. As discussed before, the elements such as chromium, molybdenum and nickel have an important role in inhibiting corrosion process. The AL-6XNPLUS™ alloy did not suffer pitting corrosion in the pressurized tests. This alloy has great amount of Cr, Mo and Ni. These alloying elements acted inhibiting the action of elements such as oxygen and chloride protecting the surface of the alloy. As discussed before, the element molybdenum has the ability to combine with oxygen and chloride forming complex oxides and salts inhibiting the action of Cl^-.

4 CONCLUSIONS

In all tests, the AL-6XNPLUS™ steel was the most resistant steel and the 316L was the least resistant steel. The type of corrosion found in the cyclic polarization tests and in the pressurized tests was pitting corrosion. The 316L had the least CO_2 corrosion resistance. This can be seen on its polarization curve. There was a formation of a protective layer but the same was not so efficient. Although 317L is considered as a conventional austenitic stainless steel, it presented a good performance in the electrochemical and in the pressurized experiments. The pressurized tests with synthetic air and carbon dioxide gases caused corrosion only for conventional austenitic stainless steels (316L and 317L). The pits on the 316L steel are 7.7 deeper than on the 317L steel. All the pits were found inside the droplets indicating that, when there was moisture present on the surface of the samples, the gases were more aggressive corroding the samples. This could not be possible only with the action of the gases by themselves. The AL-6XNPLUS™ austenitic stainless steel exhibited a good corrosion resistance. This material, although more expensive, can be the solution for use in aggressive media where there is pressure of gases and moisture acting at the same time. Conventional austenitic steels as 316L steel are not recommended for these purposes due to their low corrosion resistance in CO_2 containing environments.

5   ACKNOWLEDGMENTS

The authors would like to acknowledge the Institute of Materials Engineering of the Technische Universität Bergakademie Freiberg in Germany where part of the experiments were carried out, to the Federal University of Ceará (UFC) in Brazil, to the Brazilian Synchrotron Light Laboratory where the metallurgical characterization was carried out, to CNPq (Brazilian National Council for Scientific and Technological Development) and CAPES (Coordination of Improvement of Higher Level Personnel) for the financial support of this research in Germany and in Brazil, respectively.

6 REFERENCES

1. Idel Waisberg, Brazil's Pre-Salt Layer,
   http://large.stanford.edu/courses/2011/ph240/waisberg1/ accessed 03.02.18)

2. Anselmo N, May JE, Mariano NA, Nascente PAP, Kuri SE. Corrosion behavior of supermartensitic stainless steel in aerated and CO2-saturated synthetic seawater. Materials Science and Engineering: A. 2006; 428 (1-2): 73-79.

3. Kondo K, Ueda M, Ogawa K, Amaya H, Hirata H, Takabe H, Myazak Y. Alloy design of super 13 Cr martensitic stainless steel (development of super 13 Cr martensitic stainless steel for line pipe-1). Supermartensitic Stainless Steels'99. 1999; 11-18.

4. Russick EM, Edward M, Poulter GA, Adkins CL, Sorensen NR. Corrosive effects of supercritical carbon dioxide and cosolvents on metals. The Journal of Supercritical Fluids. 1996; 9.1: 43-50.

5. Furukawa T, Yoshiyuki I, Masanori A. Corrosion behavior of FBR structural materials in high temperature supercritical carbon dioxide. Journal of Power and Energy Systems. 2010; 4.1, 252-261.

6. Firouzdor V, Sridharan, K, Cao G, Anderson M, Allen TR. Corrosion of a stainless steel and nickel-based alloys in high temperature supercritical carbon dioxide environment. Corrosion Science. 2013; 69: 281-291.

7. Lizhen T, Anderson M, Taylor D, Allen TR. Corrosion of austenitic and ferritic-martensitic steels exposed to supercritical carbon dioxide. Corrosion Science. 2011; 53: 3273-3280.

8. Cao G, Firouzdor V, Sridharan K, Anderson M, Allen T. Corrosion of austenitic alloys in high temperature supercritical carbon dioxide. Corrosion Science. 2012; 60: 246-255.

9. Cardoso JL, Nunes Cavalcante ALS, Araújo Vieira RC, Gomes da Silva MJ, de Lima-Neto P. Pitting corrosion resistance of austenitic and superaustenitic stainless steels in aqueous medium of NaCl and H 2 SO 4. Journal of Materials Research. 2016; 31: 1755-1763.

10. Bandy R, Cahoon JR. Effect of composition on the electrochemical behavior of austenitic stainless steel in Ringer's solution. Corrosion. 1977; 33: 204-208.

11. Fontana MG. Corrosion Engineering. McGraw Hill: New York; 1986

12. Malik AU, Siddiqi NA, Ahmad S, Andijani IN. The effect of dominant alloy additions on the corrosion behavior of some conventional and high alloy stainless steels in seawater. Corrosion science. 1995; 37.10: 1521-1535.

13. AL 6XN PLUS™ Alloy Technical Data Blue Sheet, Allegheny-Lundlum corporation, 2002, http://www.alleghenyludlum.com (accessed 05.02.18)

14. Horvath J, Uhlig HH. Critical potentials for pitting corrosion of Ni, Cr-Ni, Cr-Fe, and related stainless steels. Journal of the Electrochemical society. 1968; 115.8: 791-795.

15. Willenbruch RD, Clayton CR, Oversluizen M, Kim D, Lu YC. An XPS and electrochemical study of the influence of molybdenum and nitrogen on the passivity of austenitic stainless steel. Corrosion Science. 1990;31:179 – 190

16. Sugimoto K, Sawada Y. The role of molybdenum additions to austenitic stainless steels in the inhibition of pitting in acid chloride solutions. Corrosion Science. 1977; 17.5: 425-445.

17. Sedriks AJ, Corrosion of stainless steels: 2nd ed, New York: John Wiley & Sons Inc; 1996.

18. Nordsveen M, Nes S, Nyborg R,Stangeland R. A Mechanistic Model for Carbon Dioxide Corrosion of Mild Steel in the Presence of Protective Iron Carbonate Films—Part 1: Theory. Corrosion. 2003; 59, 444-456.

19. Dostal V, Hejzlar P, Driscoll, MJ. The supercritical carbon dioxide power cycle: comparison to other advanced power cycles. Nuclear technology. 2006; 154: 283-301.

20. Loto, RT. Pitting corrosion evaluation of austenitic stainless steel type 304 in acid chloride media. Journal of Materials and Environmental Science: 2013; 4: 448-459.

21. Fierro G, Ingo GM, Mancia F. XPS investigation on the corrosion behavior of 13Cr-martensitic stainless steel in CO_2-H_2S-Cl^- environments. Corrosion. 1989; 45.10: 814-823.

22. Charng T, Lansing F. Review of corrosion causes and corrosion control in a technical facility. NASA Technical Report, TDA Progress Report. 1982; 42-69.

23. Rothman SJ, Nowicki LJ, Murch GE. Self-diffusion in austenitic Fe-Cr-Ni alloys. Journal of Physics F: Metal Physics. 1980; 10.3: 383.

Publisher: Eliva Press SRL

Email: info@elivapress.com